INCREDIBLY DISGUSTING ENVIRONMENTS™

THE HOLE IN THE OZONE LAYER AND YOUR SKIN

Kristi Lew

rosen publishing's
rosen central®

Published in 2013 by The Rosen Publishing Group, Inc.
29 East 21st Street, New York, NY 10010

First Edition

Library of Congress Cataloging-in-Publication Data

Lew, Kristi.
The hole in the ozone layer and your skin/Kristi Lew.—1st ed.
p. cm.—(Incredibly disgusting environments)
Includes bibliographical references and index.
ISBN 978-1-4488-8411-7 (library binding)—
ISBN 978-1-4488-8421-6 (pbk.)—
ISBN 978-1-4488-8427-8 (6-pack)
1. Skin—Effect of radiation on. 2. Skin—Protection. 3. Ultraviolet radiation—Physiological effect 4. Solar radiation—Physiological effect. 5. Ozone layer depletion. I. Title.
RL96.L49 2013
616.5'07572—dc23

2012029049

Manufactured in the United States of America

CPSIA Compliance Information: Batch #W13YA: For further information, contact Rosen Publishing, New York, New York, at 1-800-237-9932.

CONTENTS

INTRODUCTION

Imagine waking up to a perfect, cloudless, sunny Saturday. You're off to the park! You've got a soccer ball, a beach chair, and a couple of friends. It's going to be a great day. A couple of hours later, you've tired yourself out with an energetic game of soccer. You decide to sit down for a few minutes and wake up two hours later! Your face and arms feel a little scratchy, but you don't worry about it too much. It's time to go home.

When you get home, you catch a glimpse of yourself in a mirror. Oh, boy, are you red! You've got a nasty sunburn. Now imagine if the same thing could happen to your skin in just minutes instead of hours. That's what could happen if Earth's protective ozone layer were stripped away.

The ozone layer prevents 97 percent of the sun's harmful rays, the ones that cause sunburns and skin cancer, from reaching Earth's surface. And the ozone layer is already thin in places. In fact, it's so thin above Antarctica that scientists refer to the area as the "hole" in the ozone layer.

You may be thinking that no one lives in Antarctica, so what's there to worry about? Plenty, scientists believe. Not only would a thinner than normal ozone layer cause people to get more skin cancer, it could also destroy crops, leading to worldwide famine. It may also harm the marine life that fish rely on for food, causing fish to die and further disrupting human food supplies. There's also a link between the ozone layer and increased global warming, which, in the long term, could lead to climate change and that's everyone's problem. So, what can you do about it? Read on and find out.

A relaxing day at the beach can be fun, but getting a bad sunburn is just plain painful. If Earth's protective ozone layer did not exist, living organisms would soon die.

1 THE OZONE LAYER

So what exactly is the ozone layer? The ozone layer is part of Earth's atmosphere. It is located in a layer called the stratosphere. The stratosphere is just above the layer that humans live in, which is called the troposphere. The troposphere begins at Earth's surface and extends 4 to 12 miles (6 to 19 kilometers) above it.

The stratosphere begins at the top of the troposphere and extends to about 31 miles (50 km) above the surface of the planet. The ozone layer is a layer of concentrated ozone molecules located within the stratosphere. This layer of ozone molecules provides a natural barrier to the harmful ultraviolet (UV) radiation emitted by the sun. The amount of ozone in the stratosphere varies with location and season. Most stratospheric ozone is made in the tropics and then pushed by atmospheric winds toward Earth's poles.

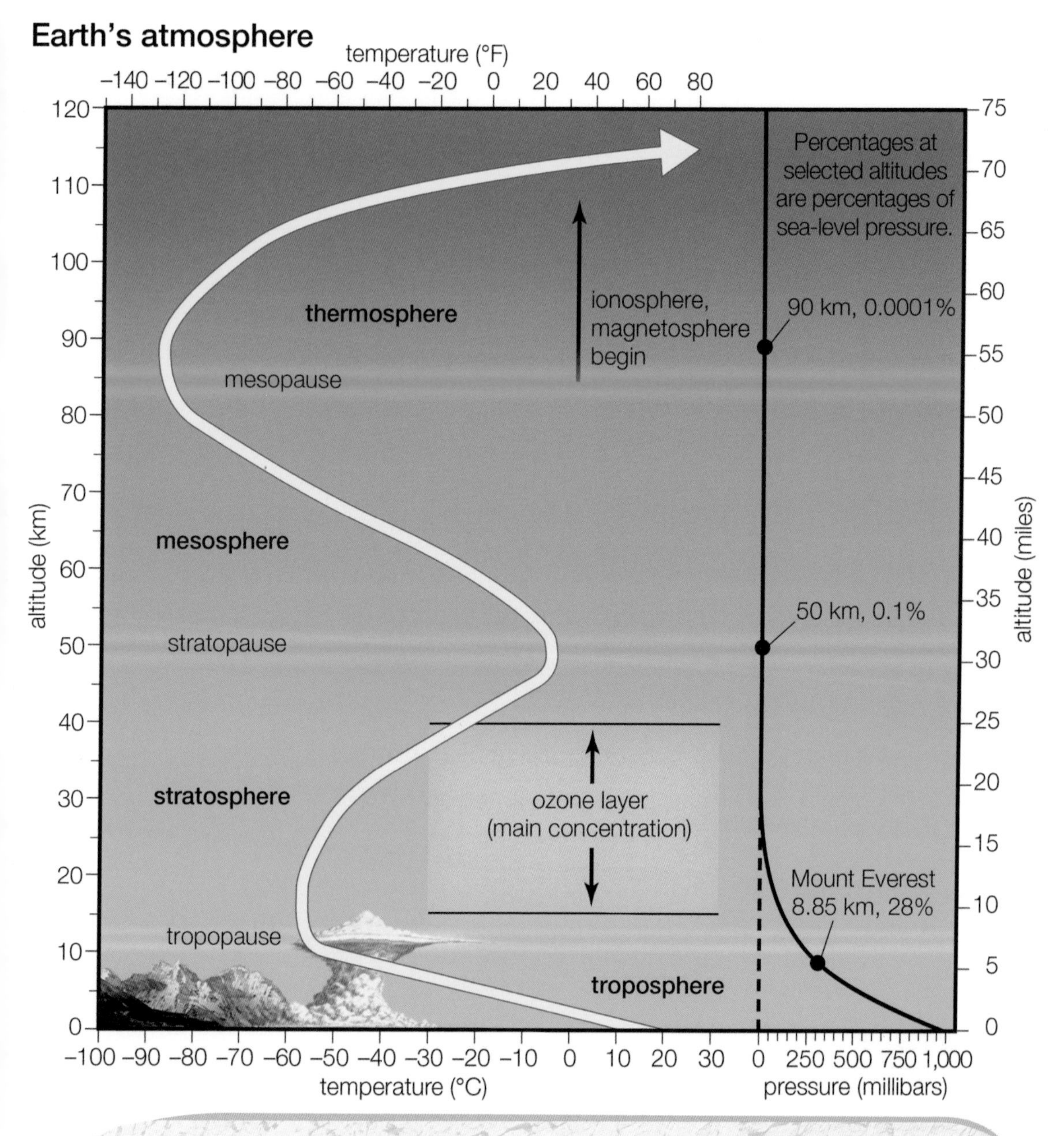

Scientists divide Earth's atmosphere into four sections. The section closest to the surface is called the troposphere. Above the troposphere is the stratosphere, which contains a layer of concentrated ozone. (The yellow arrow represents the response of temperature to increasing height.)

Ozone is a naturally occurring gas made up of three oxygen atoms chemically bonded together. Unlike the colorless, odorless oxygen people breathe, ozone is blue and smells like bleach. Ozone is much less common than oxygen. According to the Environmental Protection Agency (EPA), a sample of ten million air molecules would contain about two million molecules of oxygen and only three ozone molecules. (The other eight million or so air molecules would be mostly nitrogen.)

Ozone is constantly being made and broken down in Earth's atmosphere. When ultraviolet light hits an oxygen molecule (O_2), the molecule is broken down into two individual oxygen atoms. When one of these oxygen atoms encounters an oxygen molecule, the two can combine to form ozone (O_3).

$$O_2 + O \xrightarrow{\text{sunlight}} O_3$$

Ultraviolet light also catalyzes the reverse process and breaks ozone down into an oxygen molecule and an oxygen atom.

$$O_3 \xrightarrow{\text{sunlight}} O_2 + O$$

This natural production and breakdown of ozone is called the ozone-oxygen cycle. This cycle usually keeps the number of ozone molecules in the atmosphere in equilibrium, which means that about the same number of ozone molecules are created as are broken down. However, UV light is not the only thing that can break ozone down. Chemicals, such as chlorine and bromine can, too.

Ozone Depletion

The equilibrium between ozone creation and destruction can change because of natural processes, such as the change in seasons or volcanic eruptions, which release chemicals that break down ozone. However, human activities can disrupt this equilibrium, too.

In 1985, scientists from the British Antarctic Survey discovered something odd. They found that the ozone layer around the Antarctic was not as thick as it should be. Most scientists now

This screenshot, which is from a NASA Web site (http://www.nasa.gov/topics/earth/features/ozone-2011.html), of Earth's atmosphere over Antarctica shows the relative amounts of ozone in different colors. The areas in purple and blue contain less ozone than the red and yellow areas.

believe that human activity had caused an imbalance in the ozone-oxygen cycle. Ozone was being broken down faster than it could be replenished.

Scientists discovered that one of the main culprits in the depletion of the ozone layer was a class of chemicals called chlorofluorocarbons, or CFCs. These chemicals were once widely used as refrigerants in refrigerators and air conditioners and were found in nearly all aerosol cans.

When CFCs reach the stratosphere, ultraviolet light causes them to break down and release chlorine atoms (Cl). A chlorine atom reacts with ozone to form chlorine oxide (ClO) and oxygen molecules.

$$Cl + O_3 \xrightarrow{\text{sunlight}} ClO + O_2$$

The chlorine oxide then reacts with oxygen atoms that are produced by the ozone-oxygen cycle to produce chlorine atoms and oxygen molecules.

$$ClO + O \xrightarrow{\text{sunlight}} Cl + O_2$$

The end result is a decrease in ozone and oxygen atoms. Remember that the ozone-oxygen cycle requires oxygen atoms to react with oxygen molecules to form ozone.

Since the 1980s, Antarctica has experienced a higher ozone depletion rate than that of other places on Earth. One of the main reasons the stratosphere above Antarctica is affected so much can be attributed to a phenomenon called the Antarctic polar vortex. When the sun goes down in Antarctica in March, it marks the beginning of a long, cold winter in this region. The sun will not be seen again until September. During these months of darkness, the land and air temperatures drop dramatically. The stratosphere above Antarctica gets colder than anywhere else on Earth. Temperatures frequently dip below -112°F (-80°C).

Antarctica is also the windiest place on Earth. In May and June, high above the continent, the stratospheric winds start to build. They swirl counterclockwise in an enormous ring called the Antarctic polar vortex. CFCs become trapped in this vortex. As temperatures inside the vortex fall, water vapor and trapped CFCs condense to form tiny ice crystals. When the temperature falls below -112°F (-80°C), the ice crystals come together and concentrate in clouds called polar stratospheric clouds. All winter long, chemical reactions inside the clouds break down the CFCs, releasing chlorine molecules (Cl_2).

In its molecular form, chlorine cannot attack ozone molecules. However, in August, when the sun starts to rise over Antarctica again, the ultraviolet radiation causes the chlorine molecules to break down into chlorine atoms.

$$Cl_2 \xrightarrow{\text{sunlight}} Cl + Cl$$

As discussed above, chlorine atoms are ozone destroyers. According to the EPA, a single chlorine atom can destroy more than one hundred thousand ozone molecules. Another chemical element, called bromine (Br), which reacts similarly with ozone, can be even more destructive. The concentration of these chemicals in the polar stratospheric clouds above Antarctica resulted in a higher rate of ozone destruction in that region.

However, Antarctica is not the only place on Earth that a polar vortex and polar stratospheric clouds form. They also form over the North Pole. What's more, for the first time in the winter of 2011, scientists saw a thinning of the ozone layer above the Arctic as well as in the Antarctic. This thinning of the ozone layer above the North Pole was something new. Scientists believe that the thin

The World's Reaction

National Geographic reports that nearly 90 percent of the chlorofluorocarbons in the atmosphere were released by industrialized nations, such as the United States and European countries. In 1987, twenty-four countries signed a treaty called the Montreal Protocol, promising to reduce the consumption and to phase out the production of CFCs. Currently, the level of ozone-depleting chemicals in the stratosphere is decreasing. However, scientists believe it will take several more decades for chlorine levels to fall back to their natural levels. The World Health Organization expects that a full recovery of the ozone layer will take until at least 2050.

layer of ozone and, consequentially, an increased amount of UV radiation may have reached as far south as New York that winter.

Ultraviolet Radiation

Ozone molecules protect the people, plants, and animals on Earth by absorbing UV radiation coming from the sun. A thinner ozone layer—one with fewer ozone molecules in it than normal—allows more UV radiation to reach the surface of the planet.

Ultraviolet radiation is part of the electromagnetic spectrum. This is the same spectrum that includes visible light, radio waves, microwaves, and X-rays. All parts of the electromagnetic spectrum are types of radiation. Radiation is energy that spreads out as it travels. Electromagnetic radiation travels in waves. The waves of each type of energy in the electromagnetic spectrum—light waves, radio waves, microwaves, ultraviolet waves, and all the rest—have a characteristic shape and length. The distance between the peak of one wave and the peak of the next is called the wavelength. Gamma rays have the shortest wavelength in the electromagnetic spectrum. Radio waves have the longest.

The amount of energy produced by a type of electromagnetic radiation is inversely proportional to its wavelength. In other words, the shorter the wavelength, the more energy the particles in the wave carry. Gamma rays have the shortest wavelength. Therefore, gamma rays are the most energetic of all the electromagnetic waves. Radio waves have the longest wavelengths, so they carry the least amount of energy.

Because different types of electromagnetic radiation carry different amounts of energy, they have different biological effects. Gamma rays, for example, can do the most damage to the human body and are the primary cause of radiation sickness.

Ultraviolet radiation is generally split into three types—UVA, UVB, and UVC—based on their wavelengths. UVC has the shortest wavelength of the three and, therefore, is the most energetic. Because of the energy it carries, UVC has the potential to cause more damage to human tissue than the other two types of UV radiation. However, Earth's ozone layer filters out virtually all UVC radiation, and the air in the lower atmosphere blocks the small amount that does get through the ozone layer. Therefore, these rays normally do not reach Earth's surface.

UVA has the longest wavelength and the least amount of energy of the three types of UV radiation. The wavelength of UVB radiation lies between those of UVA and UVC. Both UVA and UVB radiation can penetrate Earth's atmosphere and reach the surface, where they can contribute to skin damage and other health issues.

2 UV RADIATION AND YOUR SKIN

UVA radiation easily passes through the ozone layer and accounts for most of the UV radiation people, plants, and animals at the surface are exposed to. While UVA radiation is harmful to humans, UVB radiation is often of more concern because it has a particularly damaging effect on DNA. The ozone layer plays a vital role in protecting humans, plants, and animals because it captures most of the UVB radiation coming from the sun. Therefore, the main concern over the depletion of the ozone layer lies in the fact that more UVB radiation could reach Earth's surface. That would leave humans, plants, and animals vulnerable to the damage that UVB radiation could do to their bodies.

Skin Cancer

The most immediate effect of overexposure to UV radiation is a painful sunburn. As uncomfortable and itchy as a serious sunburn might be, the long-term effects of sunburned skin are even more alarming. Some of the health problems caused by overexposure to UV radiation take years, or even decades, to show themselves.

One of those underlying health problems is skin cancer. According to the Skin Cancer Foundation, skin cancer is the most common form of cancer in the United States. More Americans are diagnosed with skin cancer every year than with breast,

A doctor examines a patient's skin cancer; on the right is a digital image of a sample of skin cancer. If the DNA of a skin cell is damaged, the cell may grow out of control, resulting in skin cancer.

lung, prostate, and colon cancer combined. With more than 3.5 million new cases each year, one in five Americans will be diagnosed with skin cancer in their lifetime. Every hour, one American dies of skin cancer. The largest risk factor for developing skin cancer is unprotected exposure to UV radiation—especially during childhood.

Melanoma

Melanoma, also called malignant melanoma, is the most serious and deadly form of skin cancer. It is also one of the most common forms of cancer found in adolescents and young adults from ages fifteen to twenty-nine. According to the American Cancer Society, about 4 percent of the two million people diagnosed with skin cancer each year will have malignant melanoma. That may not sound like a lot, but melanoma is a particularly dangerous form of skin cancer. It is responsible for more than 75 percent of skin cancer deaths.

Getting severely sunburned during childhood is one of the major risk factors for developing melanoma later in life. Other risk factors include a family history of melanoma and a weakened immune system. Although not much can be done about a genetic predisposition or a compromised immune system, sunburns can be prevented. If there is a family history of melanoma, regular screenings by a qualified dermatologist may be a good idea, too. Catching melanoma early saves lives.

Basal Cell and Squamous Cell Carcinoma

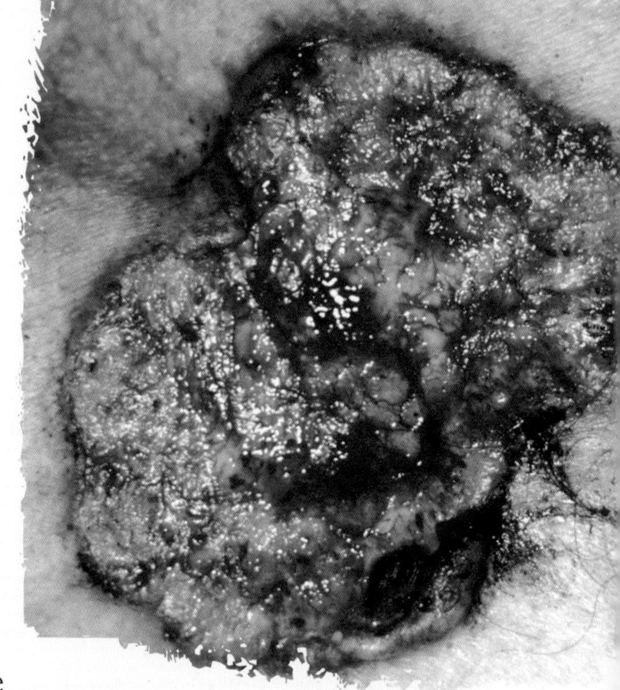

Melanoma is not the only form of skin cancer. Basal cell carcinoma (BCC) and squamous cell carcinoma (SCC) are the most common forms of nonmelanoma skin cancer. Nonmelanoma skin cancers are generally less deadly, but they can still grow, spread, and cause debilitating health problems if they are not treated promptly. The Skin Cancer Foundation estimates that 40 to 50 percent of Americans who live to be sixty-five years or older will likely develop one of these types of skin cancer at least once.

Basal cell carcinoma is the most common type of skin cancer. BCCs may look like an open sore or a red, irritated area. Pink growths, shiny bumps, or scarlike areas could also be basal cell carcinomas. The small, fleshy tumors are most often seen on areas

A red, irritated area or sore that does not appear to be healing on its own could be basal cell carcinoma (BCC) and should be checked by a doctor. Most BCC cancers can be cured if they are diagnosed and treated early.

that are commonly exposed to the sun, such as the head and neck. However, they can occur anywhere on the body.

Basal cell carcinomas originate in the deepest layer of the epidermis. The epidermis is the outermost layer of the skin. In some cases, BCCs can grow aggressively and invade other parts of the body, a situation that can turn deadly. When cancer spreads to other parts of the body, it is called metastasis.

Squamous cell carcinomas are the second most common form of skin cancer. SCCs originate in the skin's squamous cells, which make up most of the epidermis. The Skin Cancer Foundation reports that approximately 700,000 Americans are diagnosed with SCC each year, resulting in about 2,500 deaths. SCCs may look like an open sore, a wart, or a red patch. They are often crusty and sometimes bleed. Like BCCs, SCCs can metastasize, or spread, if they are left untreated.

What Causes Cancer?

Cancer is caused by out-of-control cell growth. The information that tells a cell when to grow and when to stop is carried in the cell's deoxyribonucleic acid (DNA). Exposure to UV radiation and other carcinogens can damage DNA. The more damage done to a cell's DNA, the more likely cancer is to develop. This is the reason people are more likely to develop cancer as they grow older—there is more time for DNA damage to build up. This is also the reason doctors pay particular attention to a family history

The UV Index

The UV index is a scale developed by the National Weather Service and the EPA. Calculated by ZIP code, the UV index takes into account local cloud cover and other weather conditions in order to predict how much UV radiation will make it to the ground in that particular area. Forecasters use this information to estimate people's risk of overexposure to the sun's rays on a particular day.

UV Index Range	Exposure Risk
2 and below	low
3 to 5	moderate
6 to 8	high
8 to 10	very high
11 and higher	extreme

The UV index range is issued daily and can help people make smart choices about being out in the sun.

of cancer. People in the same family may carry the same DNA structure. Certain changes in DNA structure may make a person's risk of developing cancer higher than that of people with a different DNA structure. This is called a genetic predisposition.

Solar Keratosis

Skin cancer is not the only consequence of overexposure to UV radiation. Small, rough, raised patches of skin called solar, or actinic, keratosis may also be a problem. Solar keratosis normally occurs on areas of the body that are exposed to the sun for long periods of time, such as the face, scalp, back of the hands, or upper part of the chest. A solar keratosis generally starts out as a flat, scaly area of skin that may be gray, red, pink, or skin-colored. The area may later develop into a wartlike bump with a white or yellow crusty top. These bumpy skin patches can turn into squamous cell carcinomas. Therefore, doctors consider solar keratosis a precancerous, or a premalignant, condition.

Because a solar keratosis can turn into skin cancer, a doctor should examine any suspicious-looking skin bumps. Doctors treat solar keratosis in a number of different ways. The bumps may be burned or frozen off of the skin. The abnormal cells can also be scraped off using a metal probe heated with electricity in a process called electrocautery. The growths may be cut out of the skin, too. In January 2012, the U.S. Food and Drug Administration (FDA) also approved a new medicated gel that can eliminate actinic keratosis lesions in two to three days.

People with fair skin, blue or green eyes, and red or blonde hair are more likely to develop solar keratosis than people with darker complexions. Like skin cancer, people who spend a lot of time in the sun and those who suffered many severe sunburns early in life are also more at risk for developing solar keratosis. Age is also a factor. The best way to prevent solar keratosis later in life is to protect yourself from the sun's rays now.

Premature Aging

Another common consequence of repeated sun exposure is leathery, sagging skin. Signs of premature aging include wrinkles, pigmentation changes, and loss of elasticity in the skin. Exposure to UV radiation can break down the collagen and elastin in the skin. These two chemicals keep skin smooth and young looking. When they break down, wrinkled, sagging, leathery skin is often the result. UV radiation can also cause dark spots on skin that is exposed to the sun. These changes in pigmentation are often called sunspots or age spots.

How to Protect Yourself

Lying out at the beach or in the backyard can be fun and relaxing, but too much of a good thing can be dangerous, too. Because of the damage it can do, it is always best to protect your skin and eyes from the harmful effects of UV radiation. By following a few simple steps, you can still have fun in the sun while protecting yourself from overexposure.

- Use sunscreen—even on cloudy days. The American Academy of Dermatology recommends that everyone use a water-resistant, broad-spectrum sunscreen with a sun protection factor (SPF) of 30 or higher. Broad-spectrum sunscreens protect against both UVA and UVB radiation. Sunscreen should be applied generously at least twenty to thirty minutes before going outside. Reapply sunscreen at least every two hours—more often if swimming or sweating is involved.

- Avoid tanning beds. Lying out in the sun is not the only way skin can be exposed to UV radiation. Tanning bed lamps emit UVA and UVB radiation, too.

Research conducted by the World Health Organization's International Agency for Research on Cancer has determined that the use of tanning beds before the age of thirty can increase the risk of developing melanoma by 75 percent.

- If sunscreen is not an option, protect yourself with the proper clothing. A long-sleeved shirt and long pants help filter out UV radiation. A wide-brimmed hat and sunglasses will help protect your face and eyes.
- Seek shade often between 10:00 AM and 4:00 PM, when the sun's rays are at their strongest.
- Apply sunscreen more often at the beach and on the ski slopes. Sand, water, and snow reflect UV radiation, making sunburns more likely.
- Pay attention to the UV index for your area.

In 2011, an FDA official announced that all sunscreens must go through standardized testing. Further, under new rules, sunscreen products that are labeled as "Broad Spectrum" and "SPF 15" or higher ensure consumer protection from many forms of sun-induced skin damage.

In June 2011, the FDA announced that all over-the-counter sunscreens would be required to go through standardized testing to determine their effectiveness. In addition, as of December 2012, all sunscreen labels had to reflect the results of this testing. The new labels state if a sunscreen can protect people from UVA radiation, UVB radiation, or both UVA and UVB radiation. The new labeling tells consumers whether a sunscreen provides protection against skin cancer and premature aging or if it only protects against sunburn. To qualify for the skin cancer and premature aging protection label, a sunscreen will need to block both UVA and UVB rays and have an SPF of 15 or higher.

In addition, sunscreen manufacturers will no longer be able to use the words "waterproof" or "sweat proof." The American Academy of Dermatology feels that these labels are misleading because all sunscreens eventually wash off. If FDA testing proves that the sunscreen is water resistant, the manufacturer may say so as long as it also says how long the protection lasts while swimming or sweating.

MYTHS & FACTS

Myth: There is a big, gaping hole in the ozone layer.

Fact: In some parts of the stratosphere, particularly over the Antarctic, the ozone layer is thinner than in other places. People often mistakenly refer to this thin area as a "hole" in the ozone layer.

Myth: Antarctica is the only place where ozone depletion is occurring.

Fact: Scientists have detected ozone depletion in all latitudes outside the tropics.

Myth: There is no link between ozone depletion and higher UV levels at Earth's surface.

Fact: Experiments have shown that ozone absorbs UVB radiation. Scientists have found a clear connection between decreased ozone levels in the atmosphere and increased UVB radiation at the surface.

3 MORE CONSEQUENCES OF EXCESS UV RADIATION

Your skin is not the only thing at risk when it comes to overexposure to the sun's harmful rays. UV radiation can damage your eyes, lips, and immune system, too. And these are just the ways UV radiation directly affects your body. It can also damage crops and affect marine animals, which, in the long run, could affect human health as well. Scientists have also found that the ozone layer may be contributing to the problem of global warming, but not in the way you might think.

Health Effects Beyond Your Skin

Not only can exposure to excess UV radiation increase your risk of developing skin cancer, it can also do a lot of damage to your eyes. Research has shown that exposure to UVB radiation is linked to the development of cataracts. A cataract is a clouding of the eye's lens. The lenses on your eyes are usually transparent, like the lens on a camera. Just as a clouding of a camera lens would result in a fuzzy picture, a cataract can also affect your vision. Cataracts may cause colors to seem faded. They can also make it hard to bring objects into focus. Ultimately, if left untreated, cataracts can lead to blindness. However, doctors can treat cataracts with surgery. During the surgery, the cloudy lens is removed and an artificial lens is implanted. The best way to protect your eyes from the sun's harmful rays is to wear sunglasses that filter out UVA and UVB radiation.

Another eye problem that can be caused by too much sun is pterygium. A pterygium is a layer of thin, clear tissue that grows

A clouding of the eye's lens is called a cataract, which impairs vision and can gradually lead to blindness. Scientists have determined that UV radiation increases the possibility of certain cataracts.

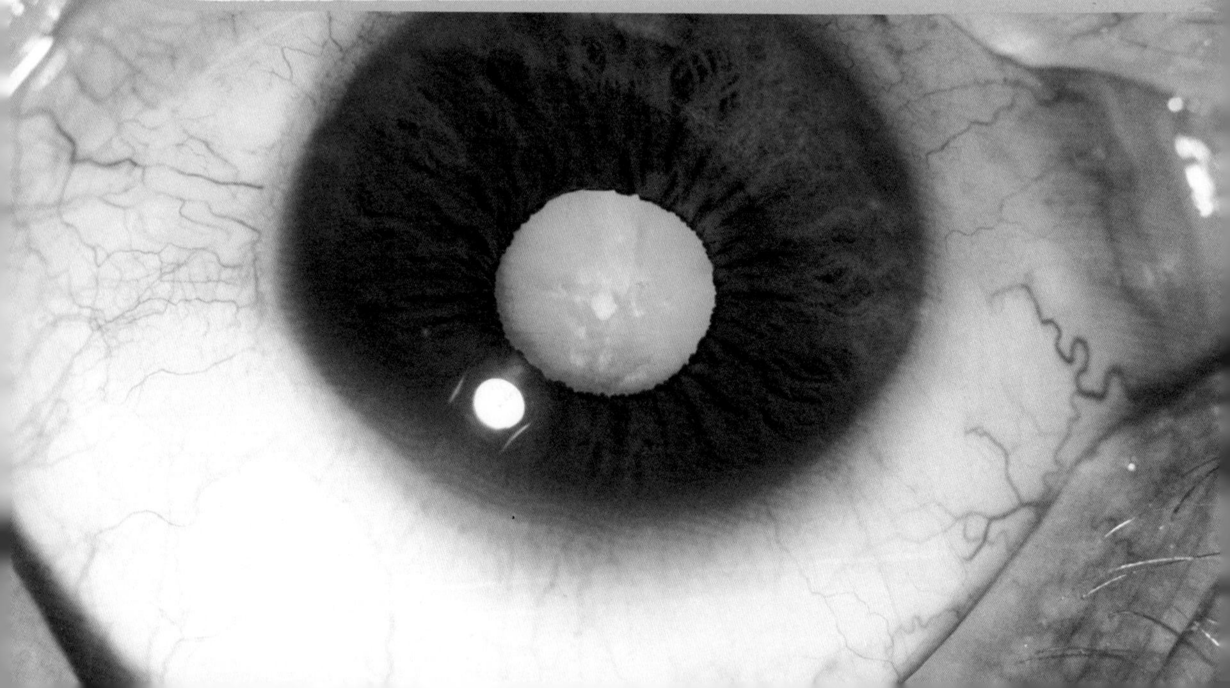

over the white part of the eye. Although a pterygium is usually painless, it can sometimes become irritated and cause an itchy, burning sensation. If left untreated, a pterygium can impair vision. A pterygium can be surgically removed, but it can also be prevented with the regular use of sunglasses that block UV rays.

Too much UV radiation can also cause a reactivation of the herpes simplex virus (HSV-1). Infection by the HSV-1 causes cold sores, or fever blisters. These small, painful blisters appear on or around the lips. A fever, sore throat, or swollen lymph nodes may accompany them, too. The HSV-1 is spread through saliva. People can become infected through kissing or by sharing toothbrushes or eating utensils with someone who has the virus. The University of Maryland Medical Center estimates that about 62 percent of Americans have been infected with the HSV-1 by the time they reach adolescence. And nearly 85 percent are infected by the time they reach sixty years old. Exposure to the sun does not cause the viral infection, but it can cause an outbreak of symptoms to occur. When you are putting on sunscreen to protect yourself from the sun's rays, do not forget to put sunscreen on your lips, too.

Scientists also suspect that overexposure to UV radiation can affect the immune system. The suppression of the immune system can leave a person vulnerable to more infections. Furthermore, those infections may be more severe than they would be if the immune system were functioning properly. A compromised immune system could also increase the risk for developing cancer.

A normally functioning immune system can often detect and destroy a cancer cell before it grows out of control. However, if too many cells are damaged by exposure to UV radiation or if the immune system can no longer destroy them, cancer cells have a better opportunity of surviving and causing a problem.

Environmental Consequences

Humans are not the only ones who are affected by increased exposure to UVB radiation. Plants are affected, too. Plants have some natural ability to reduce and repair cellular damage caused by UV radiation. They may also be able to adapt to slightly higher exposures. However, UVB radiation can change the way plants develop and grow. Research has shown that overexposure to UVB radiation can lead to the reduction in a plant's ability to defend itself against insects and other organisms that prey on it. It can also change how nutrients are distributed within the plant. Either of these changes may influence the amounts and the types of crops farmers can grow successfully. Too much UV radiation could lead to crop damage and, ultimately, to crop failure. Widespread crop failure would be disastrous for the human population as it could disrupt the food supply and lead to worldwide famine.

Land plants are not the only photosynthesizing entities to be affected by overexposure to UVB radiation. Excess UVB radiation reaching the surface of the world's oceans can also slow the

reproductive cycle of phytoplankton. Phytoplankton are made up of some types of bacteria, green algae, and other single-celled, plant-like organisms. All phytoplankton can photosynthesize, or produce their own food using the energy from the sun and gases from the atmosphere. These tiny creatures live in salt water and freshwater and form the foundation of many aquatic food webs. Without them, larger animals that feed on them, such as many species of fish, go hungry and die. In turn, animals further up the food chain do not get enough to eat. Scientists have also observed disruptions

Excess UV radiation alters the way plants grow. These tomato plants reveal the effects of exposure to different amounts of gamma radiation. The plant at the far left received no radiation. The plant at the far right had the most radiation and shows little growth.

in the reproductive cycles and early development of shrimp, crabs, frogs, and salamanders due to excess UVB radiation exposure. All these aquatic creatures play a part in the human food supply, too.

The Ozone Layer and Global Warming

While halting the destruction of the ozone layer has long been a scientific goal, scientists at the University of Leeds in England have discovered that there may be some unintended consequences. You may have heard of global warming in the news. Global warming is the increase in the average temperature of the atmosphere near Earth's surface. It is primarily caused by an increase in greenhouse

Ozone's Dual Nature

Although more than 90 percent of Earth's ozone is found in the stratosphere, where it can be beneficial to humans, the remaining 10 percent is found at ground level. Ground-level ozone is considered an atmospheric pollutant. It is harmful to humans and contributes to health problems such as asthma and other respiratory issues. At ground level, ozone is also a greenhouse gas.

gases such as carbon dioxide and methane. Ground-level ozone and CFCs are also greenhouse gases.

Sunlight warms Earth's surface, and the land and water radiate heat back into the atmosphere. Greenhouse gases absorb the heat and prevent it from escaping back into space, gradually warming the atmosphere. Earth needs some greenhouse gases to remain at a temperature that can sustain life. However, excess greenhouse gases can cause global temperatures to rise. Ultimately, scientists are concerned that global warming will change the climate patterns around the world affecting temperatures and rain patterns.

Burning fossil fuels, such as gasoline, coal, and oil, to produce energy creates carbon dioxide and other greenhouse gases.

In 2010, the scientists at the University of Leeds concluded that the thin layer of ozone over Antarctica may have actually shielded that area from the effects of global warming over the last several decades. They found that the super-fast winds that form below the thin ozone layer kick up salt spray from the ocean below. This salt spray creates moist, bright clouds that act as a mirror, reflecting the sun's rays and keeping the area cooler than it would be if these clouds did not exist. As the ozone layer heals, however, the scientists believe this effect will be reversed and increased warming will be seen in the region. This could cause a shift in global climate patterns. Scientists believe that these shifts may increase the frequency and severity of storms, droughts, and floods, which can pose an immediate danger to human health. It may also affect future health as pests that carry infectious diseases colonize new areas and air pollution increases with rising temperatures.

Another study, published in the July 26, 2012, issue of the journal *Science*, raises even more concerns about the link between global warming and the ozone layer. In this study, researchers from Harvard University report that strong summer storms can send water vapor high into the atmosphere. Once in the stratosphere, this water vapor contributes to the formation of ozone-destroying compounds. If global warming continues (and scientists believe that it will), these severe, warm-weather storms could occur more often, putting the ozone layer in further danger. Research into the link between global warming and ozone depletion is ongoing.

10 GREAT QUESTIONS TO ASK AN ENVIRONMENTALIST

1. How do we know that human actions instead of natural processes are responsible for ozone depletion?

2. What is the connection between the ozone layer and global warming?

3. Is the ozone layer still thinning?

4. Are the steps that have been taken to reduce CFC levels working and how is this affecting the ozone layer?

5. What do skeptics say about the link between human activity and the depletion of the ozone layer?

6. What chemicals have replaced CFCs? Are they safe?

7. Since the ozone layer is thinning at the North and South Poles, do people who live near the equator need to worry about increased UVB exposure?

8. Is the ozone layer likely to thin in other places other than the North and South Poles?

9. How can I help reduce ozone depletion?

10. Where can I learn more about the ozone layer?

4 HOW CAN YOU HELP?

There are several ways in which you can help prevent further damage to the ozone layer in particular and the environment in general. One of the main ways to protect the ozone layer is to prevent the release of more chemicals that harm it. Although the Montreal Protocol greatly reduced the number of products made with CFCs in many countries, products made in countries that didn't sign the treaty may still contain harmful ingredients. You may be able to find some products that advertise that they are "CFC free" or "ozone friendly." For others you may have to read the ingredients list. Try to avoid products that contain CFCs, haloalkanes, carbon tetrachloride (CCl_4), methyl chloroform (CH_3CCl_3), or hydrochlorofluorocarbons (HCFCs). Products that may contain these ingredients include, but are not limited to, industrial-strength cleaning products, fire extinguishers, glues, and aerosol sprays.

Another way to prevent harmful chemicals from being released into the atmosphere is to make sure that any repairs to your home air-conditioning system or refrigerator are done by a licensed professional. Professional service of these items should include the capture of any refrigerant released during the repair. If the refrigerator or air-conditioning system is beyond repair and needs to be replaced, also make sure the old one is disposed of and recycled properly. If your community does not have a refrigerant recovery and recycling program, you could help by starting one.

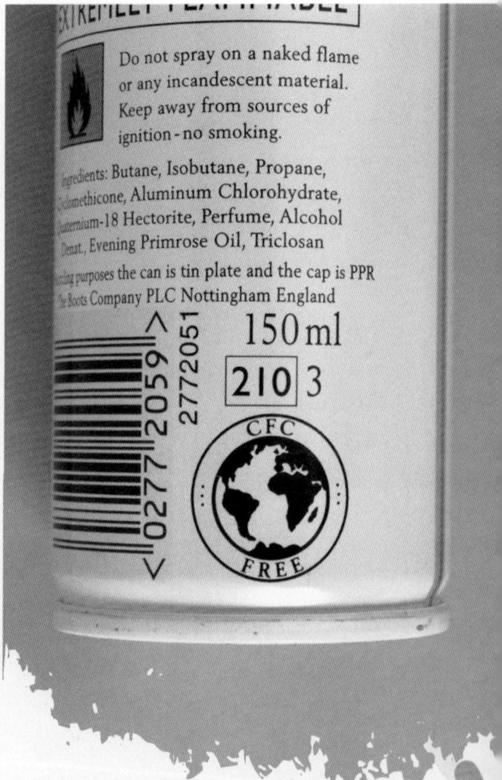

Your home air conditioner is not the only one that needs to be maintained and repaired by a professional. Automobile air conditioners should also be regularly checked for leaks to ensure no CFCs are being released into the atmosphere. If your automobile air conditioner is in need of a major overhaul, ask the repair professional about substitute refrigerants. These refrigerants do not contain CFCs and will not harm the ozone layer even if they are accidentally released. Most modern car air conditioners already contain these substitute chemicals.

Check the labels on the products your family buys. When possible, choose products advertised as "CFC free" or "ozone friendly."

Haloalkanes are hydrocarbons with one or more of the hydrogen atoms replaced by a bromine or chlorine atom. These chemicals are also harmful to the ozone layer. One major commercial use of haloalkanes is in fire extinguishers, where they are sometimes called halons. Check the fire extinguishers in your home and school. If they are haloalkane extinguishers, ask that they be replaced with another type of fire extinguisher, such as carbon dioxide or foam extinguishers.

Vitamin D

Exposure to UVB radiation does have an upside. When sunlight hits your skin, a chemical reaction in your skin cells produces vitamin D. Your body needs vitamin D to absorb calcium properly. That is why most of the milk sold in the United States is fortified with the vitamin.

Research has shown that vitamin D helps protect people from diseases such as osteoporosis, heart disease, and breast and colon cancers. In recent years, doctors have seen an increase in vitamin D deficiency in some populations. They believe decreased outdoor activity might be the cause.

But your body does not need much exposure to the sun to make all the vitamin D it needs. According to the *Harvard Public Health Review*, for example, fifteen minutes or less will do it. You can also get the vitamin D you need from fortified milk or fatty fish, such as herring or sardines.

What You Can Do About Global Warming

Global warming may put the recovery of the ozone layer in jeopardy. Greenhouse gases can come from a variety of sources. Some, like volcanic eruptions, are natural. Others are human induced.

Ground-level ozone is created in the lower atmosphere through chemical reactions between nitrogen oxides (NO_x) and volatile organic compounds (VOCs) in the presence of sunlight.

By signing the Montreal Protocol, the world community united to combat ozone depletion. Members of your community, like these teens who tend an organic garden, can also help the environment by working together to protect it from the threats facing it today.

The nitrogen oxides and VOCs are released into the atmosphere by electric power plants, car exhaust, and gasoline vapors. Carbon dioxide, another greenhouse gas, is also released in the process of burning fossil fuels in electric power plants and automobiles.

Since fossil fuels are burned to generate electricity, one of the best things you can do to decrease the amount of greenhouse gases released into the atmosphere is to conserve energy. You can conserve energy by cutting off lights, televisions, and other electronic devices when you leave the room. Taking shorter showers and running the dishwasher or washing machine only when they are full of dishes or clothes can help save energy, too. Walking or biking instead of using a car can also help cut down on the amount of gasoline that is burned.

You can also help the environment by educating yourself and others. Continue reading about environmental issues and then suggest ways you can entertain and educate your fellow students to your principal or teachers. You could also start an educational campaign in your neighborhood to make sure everyone is aware of environmental problems and how they can help. Making wise decisions that ensure the health of Earth's environment and its protective ozone layer could make a big difference to your health and to that of others.

GLOSSARY

atom The basic building block of matter; the smallest unit of a chemical element.

carcinogen A cancer-causing substance.

catalyst A substance that causes or speeds up a chemical reaction.

depletion The act of emptying or making less.

equilibrium A state of balance.

famine An extreme shortage of food.

metastasize To spread.

molecule Two or more atoms held together by covalent chemical bonds.

ozone layer A thin layer of concentrated ozone molecules located within the stratosphere.

ozone-oxygen cycle The natural production and breakdown of ozone within the atmosphere.

radiation Energy that spreads out as it travels.

stratosphere Part of Earth's atmosphere that begins at the top of the troposphere and extends to about 31 miles (50 km) above the surface of the planet.

troposphere Part of Earth's atmosphere that extends from the surface to about 4 to 12 miles (6 to 19 km) above it.

wavelength The distance from the top of one wave to the top of the next wave.

FOR MORE INFORMATION

American Cancer Society

250 Williams Street NW

Atlanta, GA 30303

(800) 227-2345

Web site: http://www.cancer.org

The American Cancer Society funds cancer research and provides information about the prevention, detection, and treatment of skin and other cancers.

Canadian Cancer Society

55 St. Clair Avenue West, Suite 300

Toronto, ON M4V 2Y7

Canada

(416) 961-7223

Web site: http://www.cancer.ca

The Canadian Cancer Society funds cancer research and provides information and support for people whose lives have been touched by cancer.

Centers for Disease Control and Prevention (CDC)

1600 Clifton Road

Atlanta, GA 30333

(800) 232-4636

Web site: http://www.cdc.gov

The CDC's Web site includes helpful information about various skin disorders and how to protect your skin.

Environment Canada

Inquiry Centre
10 Wellington, 23rd Floor
Gatineau, QC K1A 0H3
Canada
(800) 668-6767
Web site: http://www.ec.gc.ca

Environment Canada works to conserve and protect the natural environment by educating the public and developing, implementing, and enforcing environmental policies and programs for the federal government.

National Oceanic and Atmospheric Administration (NOAA)

1401 Constitution Avenue NW, Room 5128
Washington, DC 20230
(301) 713-1208
Web site: http://www.noaa.gov

The NOAA conducts research and compiles clear, concise information about the state of Earth's atmosphere and the science behind it.

U.S. Environmental Protection Agency (EPA)

Ariel Rios Building
1200 Pennsylvania Avenue NW

Washington, DC 20004
(202) 272-0167
Web site: http://www.epa.gov

The EPA provides information about environmental issues, how they affect human health, and measures that can be taken to prevent environmental harm.

U.S. Food and Drug Administration (FDA)
10903 New Hampshire Avenue
Silver Spring, MD 20993
(888) 463-6332
Web site: http://www.fda.gov

The FDA tests, regulates, and provides accurate, science-based information about food, dietary supplements, and cosmetics (including sunscreen) to help individuals make decisions about what is best for their health.

Web Sites

Due to the changing nature of Internet links, Rosen Publishing has developed an online list of Web sites related to the subject of this book. This site is updated regularly. Please use this link to access the list:

http://www.rosenlinks.com/IDE/Ozone

FOR FURTHER READING

Baker, Stuart. *Climate Change in the Antarctic.* New York, NY: Marshall Cavendish Benchmark, 2010.

Birch, Robin. *Climate Change.* New York, NY: Marshall Cavendish Benchmark, 2010.

David, Sarah. *Reducing Your Carbon Footprint at Home.* New York, NY: Rosen Publishing Group, 2009.

Gutman, Dan, ed. *Recycle This Book: 100 Top Children's Book Authors Tell You How to Go Green.* New York, NY: Yearling, 2009.

Jakab, Cheryl. *Greenhouse Gases.* New York, NY: Marshall Cavendish Children's Books, 2010.

Martins, John. *Ultraviolet Danger: Holes in the Ozone Layer.* New York, NY: Rosen Publishing Group, 2007.

Mooney, Carla. *Sunscreen for Plants.* Chicago, IL: Norwood House Press, 2010.

Morgan, Sally. *Ozone Hole.* Mankato, MN: Sea to Sea Publications, 2007.

Parker, Russ. *Climate Crisis.* New York, NY: Rosen Publishing Group, 2009.

Reilly, Carmel. *Earth's Atmosphere.* New York, NY: Marshall Cavendish Benchmark, 2012.

Stille, Darlene. *Invisible Exposure: The Science of Ultraviolet Rays.* Mankato, MN: Compass Point Books, 2010.

BIBLIOGRAPHY

American Academy of Dermatology. "Actinic Keratosis." Retrieved April 10, 2012 (http://www.aad.org/skin-conditions/dermatology-a-to-z/actinic-keratosis).

American Cancer Society. "Skin Cancer Facts." January 23, 2012. Retrieved April 10, 2012 (http://www.cancer.org/Cancer/CancerCauses/SunandUVExposure/skin-cancer-facts).

Bhanoo, Sindya. "The Ozone Hole Is Mending. Now for the 'But.'" *New York Times*, January 25, 2010. Retrieved April 10, 2012 (http://www.nytimes.com/2010/01/26/science/earth/26ozone.html?_r=1).

Skin Cancer Foundation. "Skin Cancer Information." Retrieved April 10, 2012 (http://www.skincancer.org/skin-cancer-information).

U.S. Environmental Protection Agency. "Ozone Layer Protection—Science." August 19, 2010. Retrieved April 10, 2012 (http://www.epa.gov/ozone/science/sc_fact.html).

U.S. Food and Drug Administration. "FDA Sheds Light on Sunscreens." June 14, 2011. Retrieved April 10, 2012 (http://www.fda.gov/ForConsumers/ConsumerUpdates/ucm258416.htm).

World Health Organization. "Ultraviolet Radiation." Retrieved April 10, 2012 (http://www.who.int/uv/en).

INDEX

About the Author

Kristi Lew is the author of more than forty books for teachers and young people. Fascinated with science from a young age, she studied biochemistry and genetics in college. When she's not writing, she enjoys sailing and kayaking in sunny St. Petersburg, Florida, where she always wears plenty of sunscreen.

Photo Credits

Cover © Dimitrije Tanaskovic/iStockphoto.com; p. 5 D. Sharon Pruitt Pink Sherbet Photography/Flickr/Getty Images; p. 7 Encyclopedia Britannica/UIG/Getty Images; p. 9 NASA; p. 16 (inset) Steve Gschmeissner/SPL/Getty Images; p. 16 (bottom) Peter Dazeley/Photographer's Choice/Getty Images; p. 18 Dr. Kenneth Greer/Visuals Unlimited/Getty Images; p. 23 PashOK/Shutterstock.com; p. 24 © AP Images; p. 28 Ralph C. Eagle, Jr./Photo Researchers/Getty Images; p. 31 Science Source/Photo Researchers, Inc.; p. 33 Hemera/Thinkstock; p. 37 Science & Society Picture Library/Getty Images; p. 39 Vicky Kasala Productions/Workbook Stock/Getty Images; graphics and textures: © iStockphoto.com/stockcam (cover, back cover, interior splatters), © iStockphoto.com/Anna Chelnokova (back cover, interior splashes), © iStockphoto.com/Dusko Jovic (back cover, pp. 26, 35 background), © iStockphoto.com/Hadel Productions (pp. 4, 12, 20, 32, 38 text borders), © iStockphoto.com/traveler1116 (caption background texture).

Designer: Nicole Russo; Editor: Kathy Kuhtz Campbell;
Photo Researcher: Karen Huang